I0393605

SANDIA REPORT

SAND2007-0969
Unlimited Release
Printed March 2007

Supersedes SAND2005-7959 dated December 2005

Guide to Preparing SAND Reports and Other Communication Products

Quick Reference Guide

Creative Arts Department

Prepared by
Sandia National Laboratories
Albuquerque, New Mexico 87185 and Livermore, California 94550

Sandia is a multiprogram laboratory operated by Sandia Corporation,
a Lockheed Martin Company, for the United States Department of Energy's
National Nuclear Security Administration under Contract DE-AC04-94AL85000.

 Sandia National Laboratories

Available to DOE and DOE contractors from
    U.S. Department of Energy
    Office of Scientific and Technical Information
    P.O. Box 62
    Oak Ridge, TN 37831

    Telephone: (865)576-8401
    Facsimile: (865)576-5728
    E-Mail: reports@adonis.osti.gov
    Online ordering: http://www.osti.gov/bridge

Available to the public from
    U.S. Department of Commerce
    National Technical Information Service
    5285 Port Royal Rd
    Springfield, VA 22161

    Telephone: (800)553-6847
    Facsimile: (703)605-6900
    E-Mail: orders@ntis fedworld.gov
    Online order: http://www ntis.gov/help/ordermethods.asp?loc=7-4-0#online

SAND2007-0969
Unlimited Release
Printed March 2007

Supersedes SAND2005-7959 Dated December 2005

Guide to Preparing SAND Reports and Other Communication Products

Quick Reference Guide

Creative Arts Department
Sandia National Laboratories
PO Box 5800
Albuquerque, New Mexico 87185-0619

Abstract

This *Quick Reference Guide* supplements the more complete *Guide to P reparing SAND Reports and Other Communication Products*. It provides limited guidance on how to prepare SAND Reports at Sandia National Laboratories. Users are directed to the in-depth guide for explanations of processes.

CONTENTS

Figures

TABLES

1. Introduction

This *Quick Reference Guide* is an abbreviated version of topics covered more fully in *Guide to Preparing SAND Reports and Other Communication Products* (WFS073820).

All the Sandia organizations involved in the Review and Approval (R&A) and publications process are available to assist you if you need help. Table 1 lists some of these contacts. The Creative Arts Department (NM site) or the Public Relations and Strategic Communications group (CA site) are good sources of authoritative information and assistance for your publications concerns. Table 2 lists some key web resources.

Table 1. Useful SNL Contacts

Department	Phone	Mail Stop
New Mexico (505)		
Classification	844-86991	0175
Classified Matter Protection and Control	844-0432	0166
Creative Arts*	844-6416	0619
Legal Intellectual Property	845-9536	0601
Printing & Publishing*	284-3475	0617
Review and Approval Desk	845-8220	0612
Technical Library Imaging Services	844-0774	0899
Technical Library Reference Services	845-8493	0899
Video Services*	844-7167	0650
*http://www-irn.sandia.gov/organization/div12000/ctr12600/dpt12610/labcomm.html		
California (925)		
Classification and Information Security	294-2202	9021
Legal Intellectual Property	294-3690	9031
Printing/Publishing	294-4555	9131
Public Relations & Strategic Communications http://www.ran.sandia.gov/CaLabCom/LabComCA.html	294-4555	9131
Review and Approval Desk	845-8220	9021
Technical Library Reference Services	294-3442	9211

Table 2. Directory of Web Resources

Common Look and Feel	http://www-irn.sandia.gov/organization/div12000/ctr12600/newctr12620/corpid.html
Cover templates	http://www-irn.sandia.gov/corpdata/corpforms/sandreportforms.htm
Creative Arts Web Site	http://www-irn.sandia.gov/organization/div12000/ctr12600/newctr12620/temps.html
Easy Printing PDF	\\snl\collaborative\printshop
SAND Report guide	*Guide to Preparing SAND Reports and Other Communication Products*
Library Submissions	1. Place in the Public Drop Zone folder under: SANDDOCS 2. By email to: SANDDOCS@SANDIA.GOV 3. Send the Web FileShare URL to: SANDDOCS@SANDIA.GOV
Formal vs. Programmatic R&A	Choosing a Review Process
R&A Electronic Application	https://cfwebprod.sandia.gov/cfdocs/RAA/templates/
Print Shop address for unlimited release files for printing	\\snl\collaborative\printshop
Print Shop	http://www-irn.sandia.gov/organization/div12000/ctr12600/dpt12630/pubpross.html
R&A Preflight Checklist	https://cfwebprod.sandia.gov/cfdocs/RAA/templates/
SAND Report Samples	http://www-irn.sandia.gov/organization/div12000/ctr12600/newctr12620/temps.html

1.1. SAND Reports and Communication Products

SAND numbers (e.g., SAND2005-1234) are used to track official releases of Sandia information. Official releases are information that reflects or represents policies, operations, and activities. SAND numbers consist of the year, the sequential number of the document, and, when appropriate, a suffix for the type of communication product. SAND numbers for the years 2000 and beyond use four digits for the year.

In the electronic applications, SAND numbers are issued and emailed to the author after the changes required as a result of R&A have been made. Table 3 lists the types of communications products and an example of the appropriately suffixed SAND number for each.

Table 3. Types of SAND Communication Products

Document Type	Suffix	Example
Abstract (brief or extended)	A	SAND2005-1112A
Conference Paper /Conference Presentation	C	SAND2005-1112C
Electronic Posting/Web Site	W	SAND2005-1112W
Journal Article	J	SAND2005-1112J
SAND Report	None	SAND98-1112
Other reports, Viewgraph/Presentation, Brochure/ Newsletter/Fact Sheet, Exhibit/Poster/Display, Video/DVD, Book/Book Chapter	P	SAND2000-1112P

SAND Reports that do not have the typical components of a SAND Report (e.g., promotional materials that may have customized covers) may be published with the "P" suffix.

1.2. Why SAND Reports?

SAND Reports are considered *official work product* under Sandia's DOE contracts. Although a "P" designation allows more flexibility in design and layout, it may not satisfy the customer's contractual requirement. The SAND Report format is very flexible once the requirements of the cover, title page, and distribution list are met.

2. Preparing a SAND Report

To write and prepare a SAND Report document (Figure 1), the following steps are suggested:

1. Select either the SAND Report sample or the OUO SAND Report sample from the Creative Arts web site for the kind of document that you are creating.
    http://www-irn.sandia.gov/organization/div12000/ctr12600/newctr12620/temps html

---

NOTE

Because adding a cover to your Word file later on might cause changes in the document format, it is wise to begin your document with the cover already included. Cover templates (front, back of front, and back) are available (Table 2).

---

2. Use the formatting in the sample document selected, or modify it using the Preferred Specifications.

3. Discuss with your Derivative Classifier any classification or sensitivity issues that might affect the handling of the material.

4. Input your text and illustrations. The Preferred Structure shows you how to organize your material. You should also incorporate required Common Look and Feel elements (Section 3.2) in promotional material. If copyrighted material is used, the author is responsible for obtaining permission to use the material. This includes pictures and text found on the Web.

NOTE

You may wish to use the services of a technical writer in creating the text. A graphic artist may be needed to create illustrations. Word processing assistance or a designer could help with the layout of the SAND Report. Contact Creative Arts (844-6416) for assistance obtaining any of these services.

---

5. You are responsible for properly crediting the work of others and obtaining permissions from publishers when appropriate. This includes illustrations taken from the Web (see *Guide to Preparing SAND Reports and Other Communication Products*, Section 6).

6. When your report is complete, you may wish to have it reviewed by a technical editor. You can do this even if you did not use a technical writer earlier. Your report will need a cover before going through Review and Approval. If you have not already created one, you can generate one from available templates (Table 2) or contact Creative Arts if you need help with your cover.

7. Determine the recipients of the report and set up a Distribution List at the end of the report. Be sure to include the Housekeeping Copies (Figure 8).

8. Use the classified or unclassified document checklists to be sure that the report is complete before and after Review and Approval.

9. Be sure that all required reviews are completed before submitting the report to Review and Approval. Mark the report according to its classification requirements as determined by your Derivative Classifier. Refer to Appendix A of *Guide to Preparing SAND Reports and Other Communication Products* (WFS073820) for marking instructions.

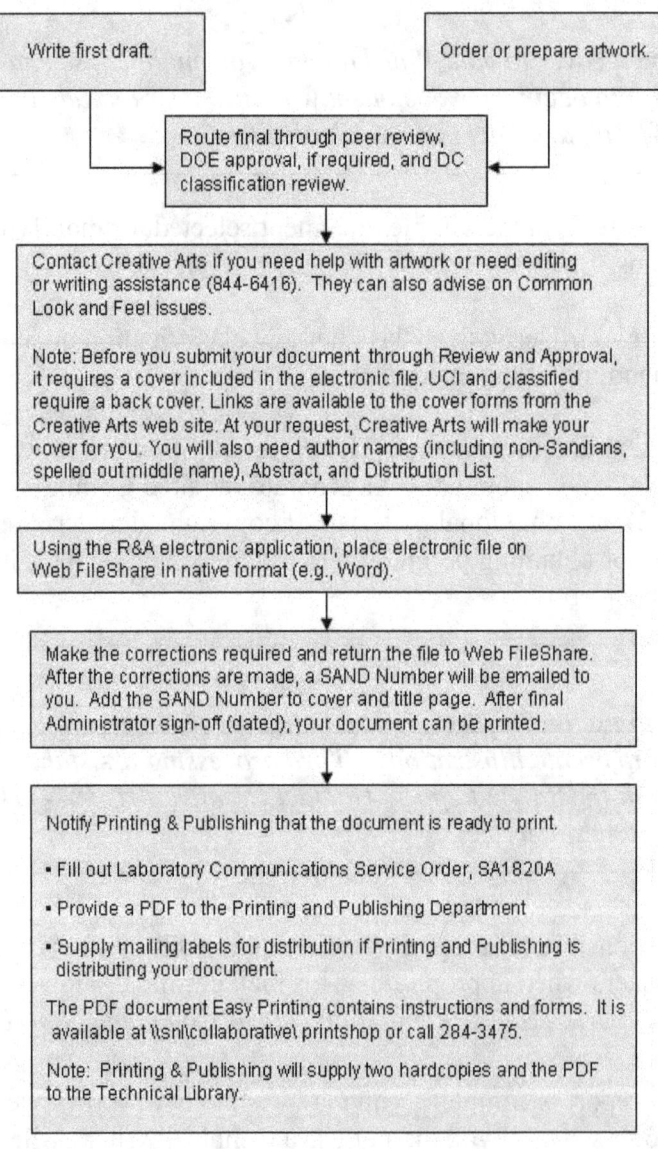

Figure 1. Flowchart for Creating SAND Reports Using Electronic Review and Approval

2.1. Unclassified Checklist

Cover
- Is all information on the cover correct? Cover should be included in the file or included separately.
- A page number does not appear on page 1 (front cover).
- Sandia address required

Check that the following cover items agree with the title page:

- Report number(s)
- Release information and print data
- Report title
- Author(s), full first name, middle initial or name spelled out (preferred), and last name of each; check spelling carefully. Authors may decide how first and middle names are presented (e.g., initials, etc.), but full first and middle names are preferred.
- Legends – Does cover include additional legends as indicated on R&A form, but not required on title page (e.g., legends for Patent Caution and Export Controlled Information)?
- Does cover have Funding Statement?
- Does the title page contain the DOE Copyright Notice (refer to *Guide to Preparing SAND Reports and Other Communication Products,* Sect. 6.1, Copyrights) for publishers other than Sandia?

Legal Notice (Disclaimer)
- Does legal notice appear on page 2?

Title Page
- Does the title page have the following information?
- Report number
- Distribution limitation (Unlimited Release, Patent Caution, or other statement as indicated on R&A form)
- Print date (month and year without comma)
- Full title of report
- Author(s), full first name, middle initial or name spelled out (preferred), and last name of each (including non-Sandians); check spelling carefully. Spelling out middle names is encouraged.
- Name of author(s)' organization(s), but not the organization number(s)
- Company name and address, including ZIP code (87185-#### for Sandia/NM). Insert author's Mail Stop in ZIP.
- The title page is numbered page 3.

Body of Report
- Is page numbering correct? Check for the following:
  - Page numbers in proper location
  - No skips in sequence
- Are the title page and the first page of body of report (the page with Introduction, Background, or similar headings) right-hand pages?
- Do Contents listings (including figures and tables) agree with the text, and are page numbers listed correctly?
- Are graphics of good, reproducible quality?
- Have you received permission to use copyrighted material (including web photos)?

Distribution

- Are names, organizations, and mail stops up to date?
- Are housekeeping copies correct?
- List one electronic copy to be sent to the Technical Library and include two print copies on the distribution for the Technical Library.

WFO Distribution list: Authors should include a Distribution list with their Work for Others report that contains at least the "housekeeping" copies.

R&A

- Your R&A is not complete until the R&A Desk dates the sign-off.
- Have you made the changes required by the R&A Desk?

REVISIONS
• The word *Revised* appears after SAND# on cover and title page *unless* a new SAND# is used.
• The supersession statement, with a blank line above it, appears below the Printed (date) line on cover and title page. The statement reads the same whether it is for SANDxxxx-xxxx Revised or a new SAND#. It is: "Supersedes SANDxxxx-xxxx dated (month & year)."
REPRINTS
• On cover, the reprint date replaces the previous print date.
• On title page, *the previous print date is retained* and the reprint date is added below it.
• On Distribution page, the original distribution is retained, followed by a blank line or two, then "Second (etc.) Printing, (date):" which is followed by another blank line and the new distribution. No housekeeping copies required unless there is a limitation or classification change.

2.2. Classified Checklist

Ref: http://www-irn.sandia.gov/iss/CPR400.3.12.1/cpr400.3.12.1.htm

Cover
- Is all the information on the cover correct?
- Sandia address required.
- Is the cover numbered page 1?

Check that the following cover items agree with the title page:

- Report number(s)
- Nuclear Weapon Data & Sigma # or other limitations
- Print date
- Report title with classification in parentheses
- Author(s), preferably first name, middle initial or name spelled out (preferred), and last name of each; check spelling carefully. Authors may decide how first and middle names are presented (e.g., initials, etc.), but full first and middle names are preferred.
- Legends
- Does cover include additional legends indicated on R&A form, but not required on title page (e.g., legends for Patent Caution and Export Controlled Information)?
- If report is going to outside recipients, does the notice "Reproduction of this document . . ." appear as last item on cover?
- Does cover have "Prepared by . . ." statement?
- Does documentation block appear on cover? Is the total number of pages correct? Page count should include Distribution list.

Legal Notice (Disclaimer)
- Does legal notice appear on page number 2?

Title Page
- Does the title page have the following information?
- Report number
- Limitation statement(s) as marked on the R&A form (such as Nuclear Weapon Data – Sigma)
- Print date (month and year without comma)
- Full title of report followed by classification [e.g., (U) or (SRD)]
- Author(s), preferably first name, middle initial or name spelled out (preferred), and last name of each; check spelling carefully
- Name of author(s)' organization(s), but not the organization number(s)
- Company name and address, including ZIP code (87185-#### for Sandia/NM)
- Does classification [e.g., (U) or (SRD)] follow the word Abstract?
- Are all classification markings and legends included?

Body of Report
- Is page numbering correct? Check for the following:
  - Page numbers in proper location
  - No skips in sequence
  - If Secret or Confidential, no double page numbers, and all blank pages state *This page Intentionally Left Blank*
- Have you received permission to use copyrighted material (including web photos)?

- Is the last page of report (which should be a blank left-hand page) numbered and does it say *This Page Intentionally Left Blank*?
- Has the back of the last page been marked properly, if required?
- Are the title page and the pages with the Introduction, Contents, Executive Summary, and similar headings right-hand pages?
- Do Contents listings (including figures and tables) agree with the text, and are page numbers listed correctly?
- Are graphics of good, reproducible quality?
- Are classified references properly marked (Section 6.3)?

Classified documents require the following:

- Page Numbers
- Subject/Title Marking
- Abstract Marking
- Classification Level
- Classification Category
- Classifier Information
- Sigmas (if applicable)
- Caveats (if applicable)
- Portion Markings (NSI only)

Distribution
- Are names, organizations, and classified mail stations current? These must be verified.
- Do all external recipients have a current classified mail channel?
- Are external recipients listed by increasing mail channel numbers?
- Are housekeeping copies correct?
- Is the Distribution page numbered?

R&A
- Does Review & Approval (R&A) Form include all approvals?

REVISIONS
• The word *Revised* appears after SAND# on cover and title page *unless* a new SAND# is used.   • The supersession statement, with a blank line above it, appears below the Printed (date) line on cover and title page. The statement reads the same whether it is for SANDxxxx-xxxx Revised or a new SAND#. It is: "Supersedes SANDxxxx-xxxx dated (month & year)."

REPRINTS
• On cover, the reprint date replaces the previous print date.   • On title page, *the previous print date is retained* and the reprint date is added below it.   • On Distribution page, the original distribution is retained, followed by a blank line or two, then "Second (etc.) Printing, (date):" which is followed by another blank line and the new distribution. No housekeeping copies required unless there is a limitation or classification change.

3. Preferred Specifications / Common Look and Feel

The sample documents on the Creative Arts Web site are modifiable and may be a useful guide on how to set up your SAND Report, including the cover:

http://www-irn.sandia.gov/organization/div12000/ctr12600/newctr12620/temps.html

3.1. Preferred Specifications

Title and Headings: Sections may be numbered, or the heading style may be varied depending upon the particular needs of the document.

Headers and Footers: Optional. May not include the Sandia logo, departmental logos, or the words "Sandia National Laboratories."

Body Text

Type Size and Font: 11- or 12-point Times New Roman or other serif type.

Line Length:
Double Column – 3 ¼ inches for each column, gutter 1/4 to 1/3 inch between columns.
Single Column – 6 inches or 6 ½ inches. Use with 10 to 12 point type.

Line Spacing: Single line spacing is preferred except for reports with equations (1.5 or double line spacing may be used).

Paragraph Style

Paragraph Indents, Spacing, and Numbering: Block or initially indented paragraphs are acceptable. Leave one or two blank lines between paragraphs. Paragraph numbering is optional.

Ragged Right Margin or Justified Right Margin: Either is acceptable. Text with a ragged right margin has been shown to be faster to read. Right-justified text may have uneven spacing that needs to be corrected.

Page Layout

Paging: Print on both sides (back to back) of the page. Headers, footers, page numbers, and margins need to alternate appropriately, depending on how you have set up the document. Centered page numbers are preferred. New sections, chapters, and appendices usually begin on an odd-numbered (right-hand) page in documents over 15 pages or in documents with many short sections. See Page Numbers below.

Page Numbers: One-half inch from bottom of page and ½ inch below last text line, centered or flush with outside margin (centered page numbers are preferred). Begin counting with the cover so that the title page is page 3. Use only Arabic numbers on SAND Reports. Unclassified documents may be numbered sequentially or by chapter or section.

For a classified report, the page number 1 shows on the front cover, and 2 appears on the back of the front cover. A page number does not appear on the front cover of an unlimited release or UCI document. Classified and UCI reports require a back cover (front and back of page); back covers are not numbered. (The Print Shop supplies an unnumbered back cover for unlimited release reports.) Classified and UCI reports that end on a right-hand (odd-numbered) page have a following blank, even-numbered page, with a page number and proper marking. Classified documents are required to have page numbers sequentially numbered.

Mark blank pages about halfway down the page with **This Page Intentionally Left Blank** only in classified documents.

Page Margins: One inch, on top, bottom, and sides. The left margin may be set at 1.25 inches. If different margins are used, the widest is the binding-side margin. Even right and left margins are preferred.

Figures and Tables

Orientation of Figures and Tables: Graphics may be presented in portrait (vertical) or landscape (horizontal) orientation, which requires the page to be turned 90 degrees counter-clockwise (to the left) to be read. Use of landscape figures is discouraged unless they are required for legibility. The caption must also be landscaped, but the page number orientation and position remains the same as for portrait pages.

Placement of Figures and Tables: Figures and tables should appear as soon as possible after they are called out in text and always within the same major section or chapter in longer documents.

Captions for Figures and Tables

Figure Captions: Should be short and end with a period, 1 to 1 ½ lines **under** the illustration. Use a font different from that of the text (e.g., 11-point Arial bold). Descriptive text after the period should be in the same font but not bolded. Block the caption under the figure unless it is very narrow, but do not run it across the page. Short captions may be centered.

Table Titles: Place 1 to 1 ½ lines **above** the column headings. They should be in the same font used for captions, ending in a period. Descriptive text after the period should be in the same font but not bolded.

Callouts: The preferred minimum type size for callouts is 10 points, although 9-point bold may be legible.

Source Lines on Figures: If a graphic is taken from another work, that information must be included with the graphic. If the graphic is taken from a non-Sandia source, obtain written permission from the previous publisher. Cite the source in the caption, or above the caption such that the source information will travel with the graphic if used elsewhere. Changing the graphic slightly does not relieve the author of the obligation to use a source line. Examples:

- "Reprinted with permission from…" if unchanged.
- "Adapted with permission from…" if slightly modified.
- "Redrawn with permission after…" if redrawn in substantially the same manner.

Oversize Figures: Oversize figures may require special handling in printing or copying. Be sure that Printing & Publishing is aware of your needs and can accommodate them.

Trademarks

The Chicago Manual of Style (15th edition, section 8.162) offers the following advice, "Although the symbols ® and ™ often accompany trademark names on product packaging and in promotional material, there is no legal requirement to use these symbols, and they should be omitted wherever possible." Including them on first use would be a recommended courtesy. Trademarks for Sandia products and those of its partners might be used more extensively.

Classified Markings

Refer to Appendix A of *Guide to Preparing SAND Reports and Other Communication Products* (WFS073820) for marking instructions. You should consult a Derivative Classifier early in the process to ensure that your drafts are properly marked and protected. Classification will review your document as part of the R&A process.

3.2. Common Look and Feel

This effort, managed by the Public Relations & Communications Center, is to convey a consistent professional image and unified corporate look on all Sandia promotional communications intended for public releases. Promotional communications products are created to attract business, promote capabilities, convey general corporate information, or generate good will. Their SAND numbers usually carry a "P" suffix, and they require Common Look and Feel elements.

NOT considered promotional materials are SAND Reports conveying scientific and technical information, project-level and program-level presentations and reports, and Web-based collaborations consisting primarily of scientific and technical works. The SAND Report cover and view graph templates include Sandia Common Look and Feel elements. Items requiring Common Look and Feel include fact sheets, fliers, brochures, displays/exhibits, video, and presentations. They are also subject to formal Review and Approval.

Before committing funds for product design, confer with one of the following to develop a design that is consistent with Common Look and Feel guidelines (they also later approve design concepts in the Formal R&A process):

Linda Lovato-Montoya (NM), MS0619, 505-844-0268
Jeff Manchester (CA), MS 9281, 925-294-2131

Approval of Common Look and Feel elements and color printing approval are required for promotional communications and are provided for through the electronic R&A application. Contact the Printing & Publishing Department (NM) or the Communication Arts Department (CA) with questions about these reviews.

Common Look and Feel elements are available at

http://www-irn.sandia.gov/organization/div12000/ctr12600/newctr12620/corpid.html

4. Preferred Structure

SAND Reports follow a standard sequence, although deviations may be made when necessary. Not all the elements appear in every document. A SAND Report **must have** a properly marked cover pages with the required statements, a title page with an abstract, and a distribution list with the housekeeping copies listed.

- Front Cover (page 1, number is not shown on unclassified report)
- Disclaimer Page (inside front cover, page 2)
- Title Page (page 3); this number appears on all reports.
- Acknowledgments (optional, page 4; this page may be left blank, but is numbered)
- Contents (page 5, or odd-numbered page)
- Preface or Foreword (optional, next blank page after Contents)
- Executive Summary (optional, odd-numbered page)
- Acronyms or Abbreviations
- Body of the Report (begins on odd-numbered page)
- References and Bibliography
- Glossary
- Appendices
- Index
- Distribution (odd or even page)
- Back Cover (UCI and Classified)

4.1. Front Cover and Disclaimer Page

The author is responsible for providing a front and/or back cover with the document (see Table 2 for templates). You may also order report covers from the Creative Arts Department (NM) or the Communication Arts Department (CA).

Note: If you use a cover template provided by, or created by, Creative Arts, you will have the required elements on your cover and meet the minimum Common Look and Feel requirements (some modification may be needed to meet the UCI and classified marking requirements).

After Review and Approval (R&A), the author inserts the SAND Number, the distribution limitation, and the expected print date. UCI and classified reports require a back cover (which is not page numbered). UCI and Classified reports require additional markings.

The Printing and Publishing Department will provide a blank back for Unlimited Release documents. You may order covers from Creative Arts, or you may create your own. Cover templates are available on the corporate forms site that include the required elements of SAND Report covers:

http://www-irn.sandia.gov/corpdata/corpforms/sandreportforms htm

The back of the front cover includes a disclaimer. Covers may be inserted in the document files or placed in a separate file for submission to Printing and Publishing. If your cover does not contain the required elements, then you should request that your SAND number assigned in R&A be suffixed with a "P." Covers are printed on preprinted Sandia stock, and the text is laid out as shown in Figures 2 and 3. In unclassified SAND Reports, front covers carry no page number. In classified SAND Reports, front covers are numbered page 1. The back of the front cover is numbered page 2.

Figure 3. Inside Front Cover (unlimited release).

Figure 2. Layout of SAND Report Cover (unlimited release).
http://www-irn.sandia.gov/corpdata/corpforms/sandreportforms.htm

4.2. Title Page

Unclassified Title Page (Figure 4)

The Title Page is numbered page 3. The information for the top block should match that on the cover. The title page may carry UCI or classified markings and legends. An Abstract is required that is approximately 150 words. Authors may decide how first and middle names are presented (e.g., initials, etc.), but full first and middle names are preferred. The names on the cover should match those on the title page.

An unclassified title page includes the following:

SAND Report Number (obtained during R&A)
Distribution Limitation (Unlimited Release, Export Controlled Information, etc.)
Print Date
Supersession Statement (if the report is a revision)
Title of Report
Authors (use middle names if available; give organization *names*).
Sandia Address
Contract or purchase order number if work was done with non-DOE funds or if report is authored
   by a non-Sandian on contract.
Abstract (unless report is being done for an agency with different requirements, e.g., NRC)
Any required notices and markings that limit distribution (such as the Non-Sandia Proprietary
   Information notice)

Classified Title Page (Figure 5)

A classified title page includes the same elements as does an unclassified title page except for the following differences:

Any special limitation such as NWD and Sigma Number
Title followed by its classification in parentheses (e.g., U or SRD)
The word Abstract with its classification in parentheses
Any required notices (such as the Restricted Data notice)

Figure 6 shows how to arrange multiple authors on a page.

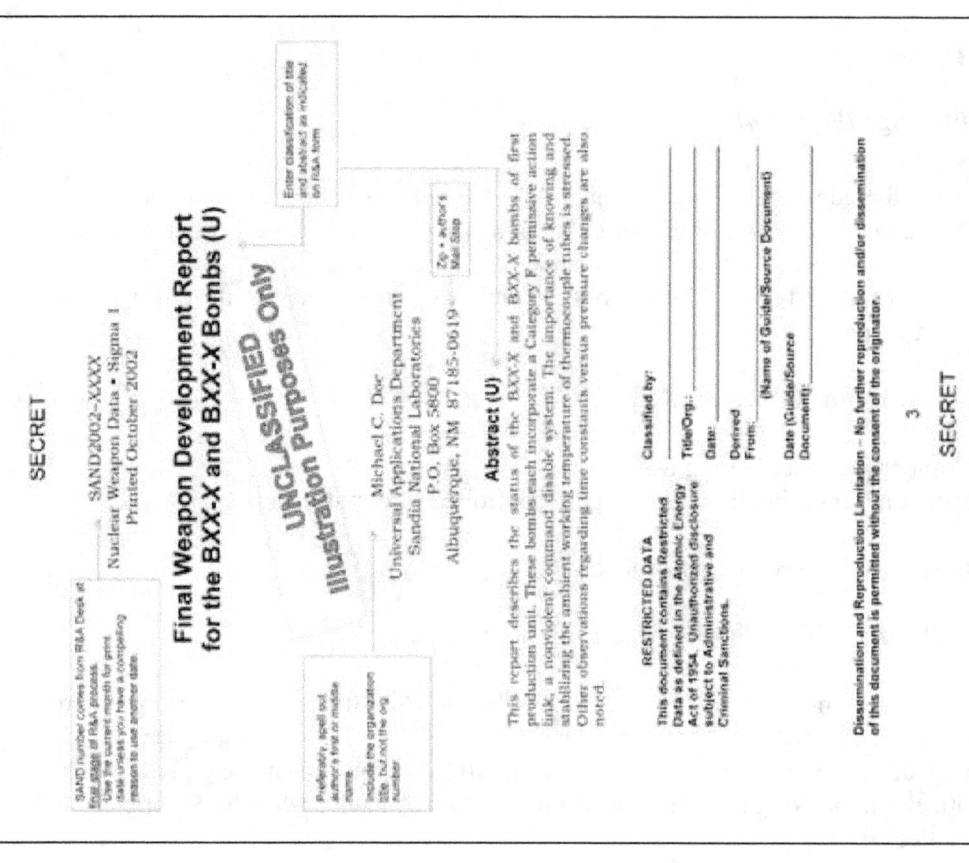

Figure 5. Classified Title Page.

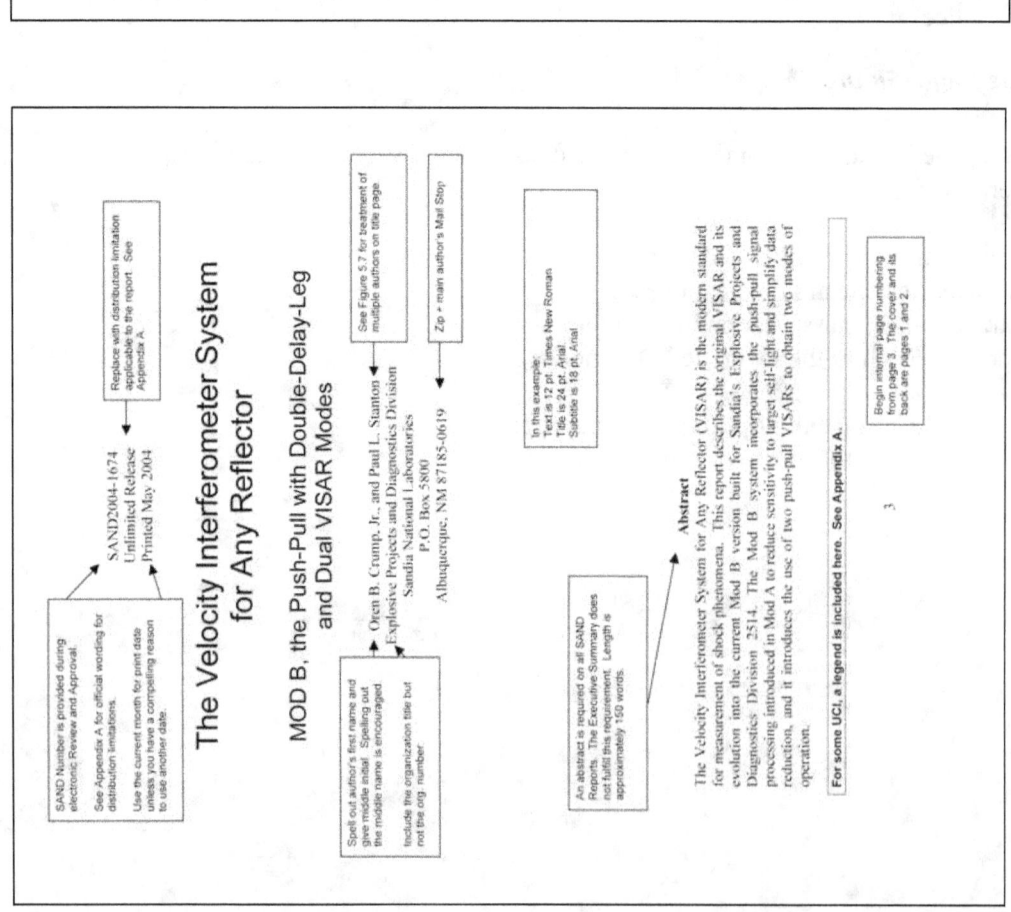

Figure 4. Unclassified Title Page.

Multiple authors:	Frank Biggs, Marion P. Apodaca, and Clarence R. Mehl Test Planning and Diagnostics Department Sandia National Laboratories P.O. Box 5800 Albuquerque, NM 87185-3415
Two or more departments:	J. William Rogers, Jr., and Stephen J. Ward Initiating and Pyrotechnic Components Department Ronald A. Guidotti Exploratory Batteries Department Sandia National Laboratories P.O. Box 5800 Albuquerque, NM 87185-2356
Sandia and an outside company:	Randall R. Nason and August E. Binder Project Engineering Department Sandia National Laboratories P.O. Box 5800 Albuquerque, NM 87185-0156 John L. Darby Science and Engineering Associates, Inc. Albuquerque, NM 87190
Editor:	Randall R. Nason, Editor Project Engineering Department Sandia National Laboratories P.O. Box 5800 Albuquerque, NM 87185-9824
Authors and editor:	Frank Biggs and Marion P. Apodaca Test Planning and Diagnostics Department Edited by Stephen J. Ward Initiating and Pyrotechnic Components Department Sandia National Laboratories P.O. Box 5800 Albuquerque, NM 87185

Figure 6. Multiple Authors on Title Page.

4.3. Acknowledgments

Acknowledgments credit substantial contributors to the work who are not authors. Contributions that are part of a person's normal job responsibilities need not be acknowledged. The acknowledgments statement usually appears on the back of the title page and is numbered page 4.

4.4. Contents

The Contents begins on an odd-numbered page, usually page 5. It should list chapters and sections by number and title, including headings in the body text through the third order (include heading numbers, if any). Back matter should be listed by number (if any) and title. It is not necessary to list the headings within the appendices (Figure 5-9). Do not list the Distribution List. A Contents section is required if the report has 15 pages or more.

If the document has more than two figures or tables, the figures and tables are listed separately by number and title. For expanded captions, give the text through the first period. Material in parentheses should be omitted unless needed to identify the figure or table.

4.5. Preface or Foreword

The Preface or Foreword is optional. It commonly begins on the next blank page after the Contents, or it may precede the Contents. It can be placed on the back of the title page if it is less than a page. The Preface includes information that is of interest to the audience but is not essential to a clear understanding of the text. A Foreword (never Forward) is written by a person other than the author after the document is finished and carries that person's name at the end. It consists of comments about the value, background, author's expertise, or other information that might add value to the report.

4.6. Executive Summary

The Executive Summary (or Summary) is optional. It begins on a right-hand (odd-numbered) page. It may also be in the body of the report as the first section. It is a self-contained, concise restatement of the major points in the body of the report. It includes information on how the work was done, the nature and purpose of the investigation, equipment and processes used, the results, and primary conclusions or recommendations.

4.7. Acronyms and Abbreviations

An Acronyms and Abbreviations list may be included if it is useful to the reader. It precedes the body of the report. The list begins on a right-hand (odd-numbered) page unless it is only one page long. A blank page follows it if it ends on an odd page (so that the body text may begin on an odd-numbered page).

Words that are not capitalized in their expanded form should not be capitalized when given in the list (e.g., IC – integrated circuit).

NOTE

Unless the list is short, words that are defined rather than expanded from the shortened form may be placed in a glossary in the back matter.

4.8. Body of the Report

The body of the report begins on a right-hand (odd-numbered) page. It may consist of chapters (shorter documents are usually divided into sections, not chapters) or sections. Numbered headings may be used, but it is desirable to limit them to two decimal places (e.g., 1.3.4). Major sections may be run together in a smaller report (few than 50 pages), but sections must begin on a right-hand (odd-numbered) page in longer reports. The report typically begins with an introduction and ends with a summary, conclusions, or recommendations section.

4.9. References and Bibliography

The References or Bibliography section follows the body of the report. References do not include items from the Appendices, which should have their own references or bibliography sections, if needed. A references section includes listings for materials quoted or referred to in the text. References are tied to particular passages in the body of the text. A bibliography may include materials consulted but not cited. A document might have either, both, or neither. Figure 7 shows two reference styles used at Sandia.

References in the text can be by number [1] for sequential references or by author/date (Hansen 1989) for author-date references.

AUTHOR-DATE

Hansen, M., 1989, *Constitution of Binary Alloys*. New Book Co., New York, NY.

Moffatt, W. G., 1994, *The Index to Binary Phase Collections*. General Electric Co., Schenectady, NY.

Nevada Department of Conservation and Mineral Resources, 1991a, *Regulations for Drilling Water Wells*. Carson City, NV.

Nevada Department of Conservation and Mineral Resources, 1991b, *Regulations and Rules of Practice for Drilling Oil and Gas Wells*. Carson City, NV, 26 p.

Oetting, F. L., M. H. Rand, and R. J. Ackerman, 1989, *The Chemical Thermodynamics of Actinide Elements and Compounds, Part I: The Actinide Elements*. International Atomic Energy Agency, Vienna, Austria.

Parr, J. G., and A. Hanson, 1995, *An Introduction to Stainless Steel*. American Society for Metals, Akron, OH.

Shunk, F. A., 1990, *Constitution of Binary Alloys, Second Supplement*. McGraw-Hill Book Co., New York.

Vol'skii, A. N., and Y. M. Sterlin, 1995, *The Metallurgy of Plutonium, Translated From Russian*. Jerusalem, Israel, Program for Scientific Translations; available from the U.S. Dept. of Commerce, Clearing-house for Federal Scientific and Technical Information, Springfield, VA.

SEQUENTIAL

1. F. L. Oetting, M. H. Rand, and R. J. Ackerman, in *The Chemical Thermodynamics of Actinide Elements and Compounds, Part I: The Actinide Elements*. International Atomic Energy Agency, Vienna, Austria, 1989.

2. M. Hansen, *Constitution of Binary Alloys*. New Book Co., New York, NY, 1990.

3. F. A. Shunk, *Constitution of Binary Alloys, Second Supplement*. McGraw-Hill Book Co., New York, 1991.

4. A. N. Vol'skii and Y. M. Sterlin, *The Metallurgy of Plutonium*, Translated From Russian. Jerusalem, Israel, Program for Scientific Translations; available from the U.S. Dept. of Commerce, Clearinghouse for Federal Scientific and Technical Information, Springfield, VA, 1995.

5. W. G. Moffatt, *The Index to Binary Phase Collections*. General Electric Co., Schenectady, NY, 1989.

6. J. G. Parr and A. Hanson, *An Introduction to Stainless Steel*. American Society for Metals, Akron, OH, 1995.

7. D. F. Bowersox and J. A. Leary, The Solubilities of Selected Elements in Liquid Pu. II. Titanium, Vanadium, Chromium, Manganese, Zirconium, Niobium, Molybdenum and Thulium, in *J Nuc Mat* Vol. 27, No. 2, pp. 181–86, February 1996.

Figure 7. Acceptable Reference Formats

4.10. Back Matter

Back Matter should be numbered or lettered in sequence following the last page of the body of the report. If chapters and sections have previously begun on right-hand (odd-numbered) pages, so should the glossary, appendices, and index.

Glossary

A Glossary is optional. The glossary defines terms that are used in the text that might be unfamiliar to some of the readers.

Appendices

Appendices are optional. Appendices support the text, but are treated separately. References in an appendix are placed in the references list for that particular appendix. They should begin on an odd-numbered page.

If an appendix is itself a publication, it may have all the components of a report, including its own contents list and page numbering. It may be necessary to reduce the pages to include the SAND Report page numbers and markings. In using previously published documents, the author is cautioned to observe copyright and distribution limitations.

Index

An index is optional. An index may be set up to update automatically when page numbers change. Otherwise, it must be constructed when the final paging of the document has been determined.

4.11. Distribution List

The last item in the report is the distribution list. Some access limitations may require a different heading for the list (check Appendix A of *Guide to Preparing SAND Reports and Other Communication Products*). The distribution list may begin on an odd- or even-numbered page, but lists with multiple pages should begin on an odd-numbered page. Do not list this section in the Table of Contents. External addresses are listed first (with any foreign addresses last). (External recipients of a classified document are arranged by increasing classified mail channels. All recipients' authorization to receive classified mail must be verified.) Then Sandia addresses are listed by increasing mail stops, followed by "record" or "housekeeping" copies retained by Sandia for its records (Figure 8).

Use Mail Stop and Org. Number. Mail sent without Mail Stops will be delayed.

UNCLASSIFIED Documents:
Unlimited Release, Limited Release, Internal Distribution. The following "housekeeping" copies should appear at the end of each SAND Report distribution list:

| 2 | MS 9960 | Central Technical Files, 8945-1 |
| 2 | MS 0899 | Technical Library, 4536 |

CLASSIFIED Documents:
List at end of **external** recipients. List **first** if there are no external recipients. For SNL/CA *classified* reports use Mail Channel and Org. number — <u>not</u> Mail Stop.

| 2 | M2497 (in **CA** use MS 9018) | Central Technical Files, 8945-1 |

(List at the end of **internal** recipients):

| 2 | MS 0899 (in **CA** use M0501) | Technical Library, 4536 |

For particular kinds of publications, add on the following as appropriate (in CA, use Classified Mail Channel number for classified submissions):

If the document has a "Patent Caution" or "Patent Interest," add the following to the END of the housekeeping copies:

| 1 | **MS 0161** | **Legal Intellectual Property, 11500** |

For CRADA documents, add the following to the END of the housekeeping copies:

| 1 | **MS 0115** | **OFA/NFE Agreements, 10112** |

For LDRD documents, add the following to the END of the housekeeping copies:

| 1 | **MS 0123** | **D. Chavez, LDRD Office, 1011** |

Figure 8. Distribution List "Housekeeping" Copies.

5. Review and Approval

Every communication product going outside Sandia, as well as internal communications with widespread distribution, is considered an official Sandia information product and requires appropriate review and approval. Even if only one Sandia employee co-authors a report with contractors, the report is considered a Sandia Report. A contractor report documents work done under a Sandia purchase order by an external entity such as a consultant or a company and is authored by that entity. The Review and Approval (R&A) Process is administered by Records Management to ensure that information released is professional, reviewed for classification, and marked appropriately.

NOTE

SAND numbers are issued after corrections identified in R&A have been approved by the manaqger. A document will not be printed until it has been signed off in R&A.

See the Choosing a Review Process Web page for detailed Review and Approval instructions (http://www-irn.sandia.gov/recordsmgmt/revapprov/5)

The two Review and Approval processes used at Sandia are the following:

Formal Review and Approval – Information released outside Sandia is assigned unique SAND numbers and must be reviewed. The R&A process and the assignment of SAND numbers are coordinated through the Recorded Information Management Department. (At Sandia/CA, Security Operations, Classifications Office, coordinates these functions.) The approval process is documented using the R&A electronic application: https://cfwebprod.sandia.gov/cfdocs/RAA/templates/

Programmatic/Organizational Review and Approval – Less formal external releases are not assigned SAND numbers and may be processed using the Programmatic/Organizational Review. These releases must be approved by the organizational management and a cognizant Derivative Classifier (DC). Before the information is released, it is recommended that the originating organization consult with the Legal Intellectual Property Center if the release is scientific and/or technical in nature and the Classification and Information Security Department if it is a potentially sensitive subject area. Documentation of the approval is also recommended using Record of Organizational or Programmatic Review for Information Release: http://www-irn.sandia.gov/corpdata/corpforms/1008rao.doc

CLASSIFIED information **MUST NOT** be uploaded into Web FileShare. The "hardcopy" routing option is used for classified requests. The submitter should fill in all the necessary information and then write the tracking number on the information release.

Warning: Do not put a classified report into the Electronic R&A System. If you are not sure about the classification level of the document, consult a Derivative Classifier or a Classification Analyst. Do not upload a classified document into Web FileShare.

Figure 9 depicts the electronic R&A process.

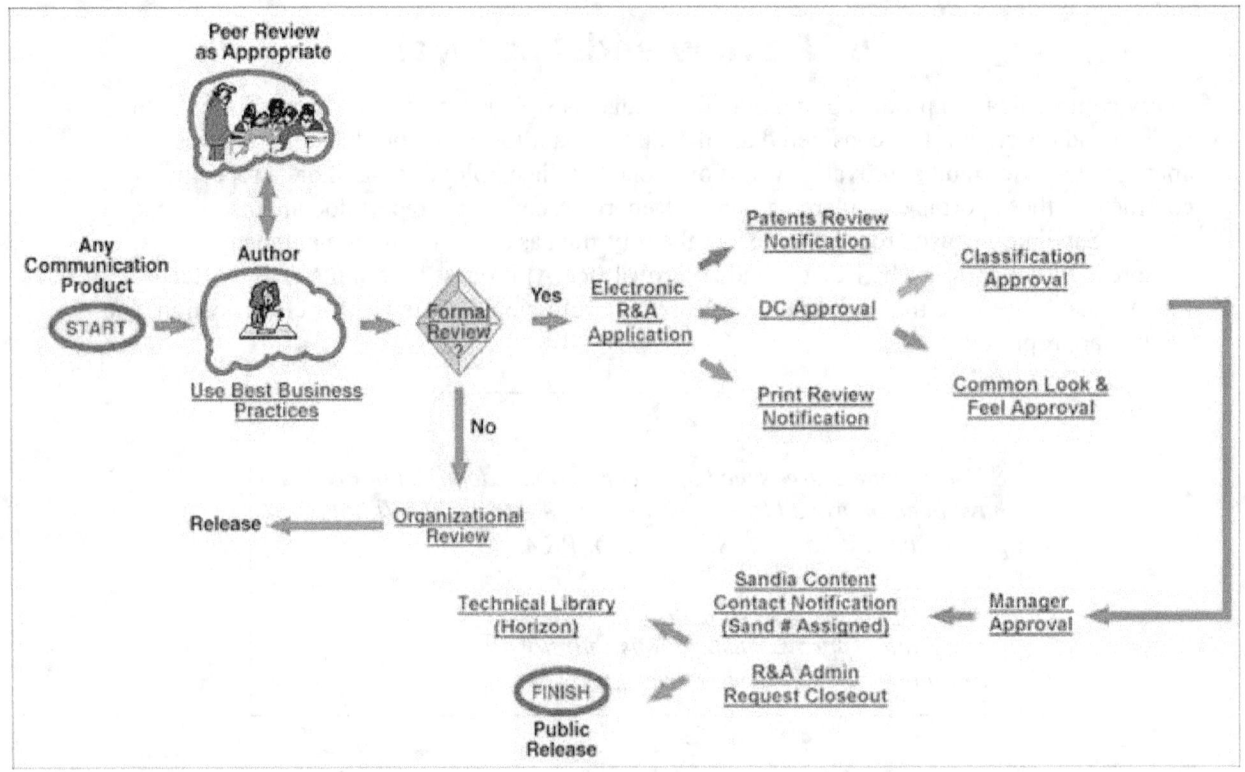

Figure 9. Flow Chart of Electronic R&A Process.

5.1. Electronic Submission of Unclassified Unlimited Release and Unclassified Limited Release

1. The author ensures that peer reviews and corrections are incorporated into the document. To ensure that the communication product is handled properly before initiation of R&A, it should be reviewed as soon as possible by a Derivative Classifier (DC) if there are any classification concerns.

2. The author obtains all necessary DOE/agency pre-approvals.

> **NOTE**
> *It is recommended that the requestor or the SNL content contact fill out the Pre-Flight Checklist prior to submitting the electronic R&A.* https://cfwebprod.sandia.gov/cfdocs/RAA/templates/

3. The author or requestor (the requestor may be someone else other than the author) uses the electronic application to initiate the R&A process:
   https://cfwebprod.sandia.gov/cfdocs/RAA/templates/

4. The requestor provides the names of authors (including non-Sandians) and all co-authors.

5. The requestor uses the electronic application to submit the R&A request and to upload the native file (e.g., Word) to Web FileShare. The complete names of all authors (including non-Sandians) and co-authors should be included.

   Promotional communications—those that promote Sandia capabilities—must be reviewed for adherence to Common Look and Feel guidelines by the Creative Arts Department (NM) or Public Relations & Strategic Communications (CA). All publications distributed outside Sandia that are printed in color or those internal publications printed in two or more inks require review.

❑ *IF* the communication product is a SAND Report (typically, a document officially released by Sandia), a cover, a title page, table of contents, and a distribution list (including housekeeping copies) are required prior to final review by the R&A Administrator (NM or CA).

❑ *IF* there is disagreement about the assigned classification sensitivity, the classification analyst will work with the DC and the author to reach an agreed-upon classification.

❑ *IF* the communication product needs revision, the reviewers will note that in the Comments field, and the author is responsible for making the changes.

6. The requestor replaces the file in Web FileShare via the electronic application if changes were made to the document.

7. Once all changes required have been agreed to by the manager, the R&A Administrator emails the SAND number to the requestor.

8. The requester downloads a cover template from the Corporate Forms web site, completes the required info, and combines it with the approved document.

   http://www-irn.sandia.gov/corpdata/corpforms/sandreportforms.htm

9. The requestor adds the SAND number to the cover and title page and replaces the file in Web FileShare.

10. The R&A Administrator indicates approval/closes out R&A. The electronic form is not closed until the Administrator block is *signed* and *dated*.

❑ *IF* the communication product is to be printed, the author contacts the NM Printing and Publishing Dept (284-3475) to send unclassified unlimited release files to the Print Shop collaborative space folder
   ❑ \\snl\collaborative\printshop (PC users)
   ❑ smb://ds09snlnt/printshop (Mac users).

   (A controlled-access UCI folder also exists in the collaborative space folder.)

   Refer to the Easy Printing PDF file at the Print Shop site for forms and information \\snl collaborative\printshop (PC). The author should carefully review all file formats that are sent to Printing and Publishing.

   http://www-irn.sandia.gov/organization/div12000/ctr12600/dpt12630/pubpross.html

11. Printing and Publishing provides two hardcopies and the PDF to the Technical Library. The Technical Library holds the record copy of all formal SAND Reports.

5.2. Hardcopy Routing (Unclassified Unlimited and Limited Release)

> **NOTE**
> *Do not use this procedure for classified material. For classified R&A, see Section 2.5, Classified Hardcopy Routing.*

❏ ***IF*** the communication product is complex video, sophisticated PowerPoint presentations with embedded video or large amounts of graphics and pictures, prepared CDs and DVDs with multiple files, or any other format not supported by Sandia's Common Operating Environment, choose "hardcopy routing" from the R&A electronic application. (The electronic application is used for tracking certain information regarding hardcopy submission of communication products.) Use the routing details page to create a routing list and record the document/tracking number on the cover of the document.

❏ ***IF*** the communication product does not fit in a folder (e.g., videos, displays), the author includes any relevant documentation (Web site map, scripts, layouts, pictures of displays, etc.) in the folder and sends the item with the folder. If that is not practical, contact the Classification and Information Security Department to determine how the item can be reviewed.

5.3 Classified Hardcopy Routing

Classified material can currently be submitted for R&A using only the hardcopy routing procedure below.

1. The requestor ensures that peer reviews are completed and corrections are incorporated into document. The Derivative Classifier should have provided an initial classification.

2. The requestor marks the material with appropriate classification and access restrictions. To ensure proper physical protection, appropriate markings and legends should be placed on the draft copy as soon as possible.

3. The requestor obtains all necessary DOE/agency pre-approvals.

4. The requestor obtains all necessary contractual or outside agency approvals.

5. It is recommended that the requestor fill out the Pre-Flight Checklist available via the electronic R&A application prior to initiating the electronic R&A. (The electronic application is used for tracking certain information regarding hardcopy submission of communication products.)

6. The requestor uses the electronic R&A application to initiate R&A. The requestor selects "hardcopy" for hardcopy routing. If the title is classified, enter "classified title" in the title

field. The requestor provides the complete names of authors (including non-Sandians) and all co-authors, with their complete middle names, if available.

7. Requestor should consult the department's Classified Administrative Specialist for logging out and handling and wrapping of classified hardcopies and/or Accountable Removable Electronic Media (ACREM).

NOTE

Use properly marked front and back cover sheets on each hard-copy of the draft document and on the master. For Secret, use Form SF 704 for the front cover and Form SF 2900-HE or -EA for the back cover; for Confidential, use Form SF 705 for the front cover and Form SA 2900-HEA for the back cover.

---

8. Put the material, with the appropriate cover sheets, in a classified folder or envelope. Staple the folder together at the top and right side to secure the contents or use strong rubber bands around the folder in both directions. Place bulky documents in an envelope.

NOTE

The requestor must hand carry the classified document package to each of the reviewers obtained through the R&A electronic application.

---

9. Attach the routing slip from the electronic application and hand carry to the Derivative Classifier, Classification and Information Sensitive Department (NM), the approving Manager, and the R&A Administrator. The reviewers approve via the electronic R&A.

❑ *IF* there is disagreement about the assigned classification, the classification analyst will work with the DC and the author to arrive at an appropriate classification.

❑ *IF* the communication product is a promotional communication (those that promote Sandia capabilities), it must be reviewed for adherence to Common Look and Feel guidelines by the Creative Arts Department (NM) or Public Relations & Strategic Communications (CA).

❑ *IF* a publication is distributed outside Sandia and is printed in color or is an internal publication printed in two or more inks, it requires review and color approval. Printing and Publishing is authorized to give this approval.

10. The requestor accesses the reviewer comments through the electronic R&A or on the hardcopy. The author makes any necessary revisions, including adding the SAND number to the title page.

11. The requester downloads a cover template from the Corporate Forms web site, completes the required info, and combines it with the approved document or includes in a separate file.

   http://www-irn.sandia.gov/corpdata/corpforms/sandreportforms htm

12. The R&A Administrator (NM) closes out the R&A processes. The R&A is not closed out until the R&A Administrator block is signed and dated. After close-out, contact Printing and

Publishing (NM) to make printing arrangements. The requestor supplies the native files and a PDF to Printing and Publishing.

Printing and Publishing provides two hardcopies and the PDF to the Technical Library. The Technical Library holds the record copy of all formal SAND reports.

NOTE

The Prime Contract requires that inventions be reported and gives the government and Lockheed Martin certain rights. It is the author's responsibility to notify Legal Intellectual Property when intellectual property might be involved.

---

6. Printing and Distribution

All printed or copied documents **must** be published through the Sandia Printing and Publishing Department (NM) and Public Relations & Strategic Communications (CA). Printing using more than one color requires DOE approval. Promotional materials require review for adherence to Common Look and Feel guidelines. Color copies do not require approval.

Once a SAND Report has been signed off through R&A, it can be printed by the Sandia Printing and Publishing Department. They prefer to receive native files as PDFs created to their specifications (see Easy Printing at \\snl\collaborative\printshop).

> **NOTE**
> *Refer to the Easy Printing PDF file, at \\snl\collaborative\ printshop for forms and information on creating the PDFs for printing. The author should carefully check all PDFs sent to Printing & Publishing. The Printing and Publishing site at \\snl\collaborative\printshop\UCI-OUO contains a folder in which UCI can be placed. (Access this site from Run on a PC.) Once you place UCI in that file, you cannot access it again.* **Classified material cannot be sent electronically to Printing and Publishing.**

Printing and Publishing can help you with distribution of copies listed in the Distribution List. They will not print more copies than are included on the Distribution List.

Make self-adhesive labels for all internal and external addressees receiving classified reports and for all internal addressees receiving Confidential or Secret reports. The Printing and Publishing Document Processing Team supplies labels for external addressees who are to receive Confidential and Secret reports. If the report is Secret, also prepare Receipts for Classified Information (RCIs) in accordance with Classified Matter Protection and Control guidance. For unclassified reports, the customer must supply mailing labels for external addressees. Printing and Publishing can print address information on the back cover of SAND Reports for unclassified internal distribution. Send the forms and labels to Document Processing.

The Technical Library is responsible for the record copy of all SAND documents (CPR400.2.13.14 Att. A). After a SAND document has been signed off through the electronic R&A application, Printing and Publishing automatically submits an electronic copy and two print copies to the Technical Library for many kinds of reports. Electronic access to SAND documents is made available through various services, depending on classification and limitations, including the Technical Library's online catalog, Sandia's search capability, and Web FileShare.

Abstracts remain on line after publication of the complete document, so updated abstracts should be sent to the Technical Library. Conference papers should also be updated if changed. Substantially changed abstracts and conference papers are subject to additional R&A. A conference paper should also include the title and date of the conference publication as well as the page numbers of the conference paper within the proceedings, if available. Notify the Technical Library whenever a Patent Caution is removed from a conference paper or journal article so that it can be released to requesters.

Appendix A: Markings and Legends

All SAND documents require a special marking or legend or both. This table lists the markings and legends to be used with various types of documents. To ensure proper physical protection, the appropriate markings and legends should be placed on the draft copy as soon as possible.

A.1 Unclassified Unlimited Release Documents

Name of Notice	Instructions	Sample
Unclassified Unlimited Release	Cover/title page – top block under SAND No.	**Unlimited Release**
	Bottom of cover	**Approved for public release; further dissemination unlimited.**

A.2 Unclassified Controlled Information (UCI)

In addition to the markings specified here, check the linked sites for other marking and protection control requirements, including special OUO/NPI cover page (SF-1008-UCA) or UCNI cover page (SF-1008-N) and the OUO/NPI envelope (SF-1008-UCB). The table entries are referenced to Classification Handbook for Derivative Classifiers at SNL or to various Sandia CPRs giving guidance on OUO:

* Classification Department Web site, Classification Handbook for Derivative Classifiers at SNL: http://www-irn.sandia.gov/security/infosec/newdoeorder/ 4225 Classification

* New Doe Order Web site (has sample covers): http://www-irn.sandia.gov/security/infosec/newdoeorder/

* Sandia CPR, Management of Official Use Only Information: http://www-irn.sandia.gov/iss/CPR400.3.3/cpr400.3.3.htm

* Sandia CPR, Sandia Management of Information Throughout Its Life Cycle: http://www-irn.sandia.gov/policy/infrastructure/cpr400220_10.htm

QUICK REFERENCE GUIDE

Name of Notice	Instructions	Sample
Naval Nuclear Propulsion Information http://www-irn.sandia.gov/security/dept/class ification/ucnimenu.htm Cover template: http://www-irn.sandia.gov/corpdata/corpform s/sandreportforms htm	Cover/title page – top block under SAND No. Cover and title page Top and bottom of each page Distribution is very limited. Note: NNPI may also apply to classified information.	**Naval Nuclear Propulsion Information** **NOFORN: This document is subject to special export controls and each transmittal to foreign governments or foreign nationals may be made only with prior approval of the Naval Sea Systems Command.** **No Foreign Dissemination** or **NOFORN**

Name of Notice	Instructions	Sample
Unclassified Controlled Nuclear Information (UCNI) http://www-irn.sandia.gov/security/dept/classification/docs/ucniBrochure.pdf UCNI box: http://www-irn.sandia.gov/security/infosec/newdoeorder/docs/UCNILegend.doc Cover template: http://www-irn.sandia.gov/corpdata/corpforms/sandreportforms htm	Cover/title page – top block under SAND No. Cover (UCNI legend) Add dissemination statement on cover and title pages. Cover, title page, and all internal pages top and bottom. Marking on interior pages may be limited to those pages containing UCNI if more convenient.	**Unclassified Controlled Nuclear Information** ● UNCLASSIFIED CONTROLLED NUCLEAR INFORMATION NOT FOR PUBLIC DISSEMINATION Unauthorized dissemination subject to civil and criminal sanctions under Section 148 of the Atomic energy Act of 1954 as amended (42 USC 2168 Review Official: <u>Name and Organization</u> Date: Guidance: (List all UCNI Guidance) **Further dissemination authorized to Department of Energy and DOE contractors only; other requests shall be approved by the originating facility or higher DOE programmatic authority.** **UNCLASSIFIED CONTROLLED NUCLEAR INFORMATION** or **UCNI** Caveat: UCNI matter may be marked with the caveat "DISSEMINATION CONTROLLED" when programmatic requirements place special dissemination or reproduction limitations on information controlled as UCNI. This marking indicates that reproduction, extraction of information, or redistribution of such matter requires the permission of the cognizant DOE program office. If this caveat is applied, the originator must ensure that the following marking is placed immediately above the matter's front legend: **DISSEMINATION CONTROLLED** **Distribution authorized to DOE and DOE contractors only.** **Other requests shall be approved by the cognizant DOE** **program office, which is _____, before release.**

Name of Notice	Instructions	Sample
Official Use Only (OUO) http://www-irn.sandia.gov/iss/CPR400.3.3/cpr400.3.3.htm Sample: http://www-irn.sandia.gov/security/infosec/newdoeorder/docs/OUOsampledoc1.doc Cover template: http://www-irn.sandia.gov/corpdata/corpforms/sandreportforms.htm	**See A.3 for subcategories.** Cover/title page – top block under SAND No. Cover (for documents to be excluded from Freedom of Information Act requests) Note: Use the organization name in the OUO box and the complete category exemption and number. Cover and title pages include the DOE dissemination statement. Place at top and bottom of each page. (Top marking is a good business practice. As a good business practice, place OUO on cover and title page. It is optional to use it only on internal pages containing OUO.)	**Official Use Only / Exemption name or subcategory (e.g., Official Use Only / Circumvention of Statute / Export Controlled Information)** --- **OFFICIAL USE ONLY** May be exempt from public release under the Freedom of Information Act (5 U.S.C. 522), exemption number and category: **Department of Energy review required before public release.** Name/Org: Date: Guidance: --- Exemption number is required. N/A 1. Circumvention of Statute 2. Statutory Exemption (Requires a subcategory) 3. Commercial/Proprietary (Requires a subcategory) 4. Privileged Information 5. Personal Privacy 6. Law Enforcement 7. Financial Institutions 8. Wells **Further dissemination authorized to the Department of Energy and DOE contractors only; other requests shall be approved by the originating facility or higher DOE programmatic authority.** **OFFICIAL USE ONLY** or **OUO**

A.3. OUO Information

Name of Notice	Instructions	Sample
Applied Technology http://www-irn.sandia.gov/iss/CPR400.3.3/cpr400.3.3.htm Cover template: http://www-irn.sandia.gov/corpdata/corpform s/sandreportforms.htm	**Include all OUO markings (refer to A.2)** Cover/title page – top block under SAND No. Cover	OUO Exemption 5, Privileged Information **Official Use Only / Applied Technology** **APPLIED TECHNOLOGY** **Any further distribution by a holder of this product or data therein to third parties representing foreign interests, foreign governments, foreign companies, and foreign subsidiaries or foreign divisions of U.S. companies shall be approved by the** *insert appropriate NE Program Office official from list below*, **U.S. Department of Energy. Further, foreign party release may require DOE approval pursuant to 10 CFR 810, and/or may be subject to Section 127 of the Atomic Energy Act.** **Further dissemination only as authorized by the originating facility or higher DOE programmatic authority; requester must possess appropriate security clearance, need-to-know, and facility approval for receipt and storage of classified documents by the DOE Office of Security Affairs.** ☐ Associate Deputy Assistant Secretary for Reactor Systems, Development, and Technology ☐ Associate Deputy Assistant Secretary for Reactor Deployment ☐ Deputy Assistant Secretary for Space and Defense Power Systems ☐ Deputy Assistant Secretary for Naval Reactors
	Mark top and bottom of each page.	**OFFICIAL USE ONLY**

QUICK REFERENCE GUIDE

Name of Notice	Instructions	Sample
Export Controlled Information (ECI) http://www-irn.sandia.gov/iss/CPR400.3.3/cpr400.3.3.htm	**Include all OUO markings (refer to A.2)**	
	Cover/title page – top block under SAND No.	**Official Use Only / Export Controlled Information**
Cover template: http://www-irn.sandia.gov/corpdata/corpforms/sandreportforms.htm	Add dissemination statement on cover and title pages.	**Further dissemination authorized to Department of Energy and DOE contractors only; other requests shall be approved by the originating facility or higher DOE programmatic authority.**
	Ane appropriate legend must be used on cover page:	**Export Controlled Information:** plus one of the following statements:
	ITAR – International Traffic and Arms Regulation	**Treat this material per Department of State (DOS) International Traffic in Arms Regulations, 22 CFR 120-130. Information contained in this document is also subject to controls defined by the Department of Defense Directive 5230.25.** http://www-irn.sandia.gov/security/infosec/newdoeorder/docs/OUOsampledocECIITAR.doc
	EAR – Export Administration Regulations	**Treat this material per Department of Commerce Export Administration Regulations, 15 CFR 730-774.** http://www-irn.sandia.gov/security/infosec/newdoeorder/docs/OUOsampledocECIEAR.doc
	DOE – Department of Energy	**Treat this material per Department of Energy Assistance to Foreign Atomic Energy Activity Regulations, 10 CFR 810.1.** http://www-irn.sandia.gov/security/infosec/newdoeorder/docs/OUOsampledocECIDOE.doc
	NRC – Nuclear Regulatory Commission	**Treat this material per the Nuclear Regulatory Commission's Export and Import of Nuclear Equipment and Material Regulation, 10 CFR 110.** http://www-irn.sandia.gov/security/infosec/newdoeorder/docs/OUOsampledocECINRC.doc
	Mark top and bottom of each page.	OFFICIAL USE ONLY

OUO Exemption 3, Statutory Exemption

QUICK REFERENCE GUIDE

Name of Notice	Instructions	Sample
Patent Caution http://www-irn.sandia.gov/iss/CPR400.3.3/cpr400.3.3.htm	**Include all OUO markings (refer to A.2)** Cover/title page – top block under SAND No.	OUO Exemption 3, Statutory Exemption **Official Use Only / Patent Caution**
Cover template: http://www-irn.sandia.gov/corpdata/corpforms/sandreportforms.htm	Cover and title page	**Further dissemination authorized to U.S. Government agencies only; other requests shall be approved by the originating facility or higher DOE programmatic authority.**
	If Patent Caution is marked on R&A form, place these legends on front cover only.	**PATENT CAUTION** **This document may reveal patentable subject matter. The information must not be divulged outside Sandia National Laboratories without the approval of the Legal Intellectual Property Center. Approved external recipients must not divulge the information to others.** **No further dissemination outside of the Government without the approval of the Legal Intellectual Property Center, Sandia National Laboratories.**
	Distribution list for a Patent Caution or Patent Interest should include:	1 0161 Legal Intellectual Property, 11500
	Mark top and bottom of each page.	**OFFICIAL USE ONLY**

Name of Notice	Instructions	Sample
(Non-Sandia Co. name) Proprietary Information http://www-irn.sandia.gov/iss/CPR400.3.3/cpr400.3.3.htm Cover template: http://www-irn.sandia.gov/corpdata/corpforms/sandreportforms.htm	**Include all OUO markings (refer to A.2)** Cover/title page/back cover top and bottom Top block under SAND No. Cover and title pages Add dissemination statement on cover and title pages. *Note: The abstract should not include any Proprietary Information.* Mark top and bottom of each page.	OUO Exemption 4, Commercial Proprietary **OFFICIAL USE ONLY** **(Non-Sandia co. name) Proprietary Information** Official Use Only / (Non-Sandia co. name) Proprietary Information **(Non-Sandia co. name) PROPRIETARY INFORMATION** **This technical data contains (Non-Sandia co. name) Proprietary Information furnished under contract or agreement (no., if applicable) between Sandia National Laboratories and (non-Sandia co. name) for the controlled release of the information. Disclosure outside the Government is not authorized without prior approval of the originator, or in accordance with provisions of 48 CFR 952.227 and 5 U.S.C. 552.** **Further dissemination authorized to U.S. Government agencies only; other requests shall be approved by the originating facility or higher DOE programmatic authority.** **OFFICIAL USE ONLY**

Name of Notice	Instructions	Sample
Protected Battery Information http://www-irn.sandia.gov/iss/CPR400.3.3/cpr400.3.3.htm	**Include all OUO markings (refer to A.2)**	OUO Exemption 3, Statutory Exemption
	Cover/title page – top block under SAND No.	**Official Use Only/Protected Battery Information**
Cover template: http://www-irn.sandia.gov/corpdata/corpforms/sandreportforms.htm	Cover only	**PROTECTED BATTERY INFORMATION** **This product contains Protected Battery Information that was produced under Contract/CRADA No. _____ and is not to be further disclosed for a period of up to five years after the completion of the individual project, or not prior to *(date)*.**
	Add dissemination statement on cover and title pages. *Note: The abstract should not include Protected Battery Information.*	**Further dissemination authorized to the Department of Energy only; other requests shall be approved by the originating facility or higher DOE programmatic authority.**
	Mark top and bottom of each page.	**OFFICIAL USE ONLY**

QUICK REFERENCE GUIDE

Name of Notice	Instructions	Sample
Protected CRADA Information http://www-irn.sandia.gov/iss/CPR400.3.3/cpr400.3.3.htm Cover template: http://www-irn.sandia.gov/corpdata/corpforms/sandreportforms.htm	**Include all OUO markings (refer to A.2)** Cover/title page – top block under SAND No. Cover and title page "Date" refers to the month and year that the most recent Protected CRADA Information was added to the report. For CRADA documents, add the following to the END of the housekeeping copies: _Note: The abstract should not include Protected CRADA Information. Mark paragraphs containing CRADA information_ Mark top and bottom of each page.	OUO Exemption 3, Statutory Exemption **Official Use Only/Protected CRADA Information** **PROTECTED CRADA INFORMATION** **This product contains Protected CRADA Information which was produced (_date_) under CRADA No. _____ and is not to be further disclosed for a period of _____ from the date it was produced except as expressly provided for in the CRADA.** **Further dissemination authorized to the Department of Energy only; other requests shall be approved by the originating facility or higher DOE programmatic authority.** 1 0115 OFA/NFE Agreements, 10112 OFFICIAL USE ONLY

Name of Notice	Instructions	Sample
Small Business Innovation Research (SBIR) and Small Business Technology Transfer (STTR) http://www-irn.sandia.gov/iss/CPR400.3.3/cpr400.3.3.htm Cover template: http://www-irn.sandia.gov/corpdata/corpforms/sandreportforms.htm	**Include all OUO markings (refer to A.2)** Cover/title page/back cover top and bottom; top block under SAND No. Add dissemination statement on cover and title pages. Mark top and bottom of each page.	OUO Exemption 3, Statutory Exemption **OFFICIAL USE ONLY / SBIR PROPRIETARY INFORMATION** OR **OFFICIAL USE ONLY / SMALL BUSINESS TECHNOLOGY TRANSFER** **Further dissemination authorized to U.S. Government agencies only; other requests shall be approved by the originating facility or higher DOE programmatic authority.** **OFFICIAL USE ONLY**

A.4. Classified Documents

Classified documents have levels of classification: TOP SECRET (TS), SECRET (S), or CONFIDENTIAL (C). They also have categories: RD, FRD, or NSI. They can also have caveats or access restrictions such as NWD, Sigma numbers, CNWDI, and NOFORN. These levels, categories, and access restrictions may be used in various combinations. Detailed guidance on handling classified documents can be found in the publications:

ACREM/CREM Marking Guide
http://www-irn.sandia.gov/security/c_docs/man/man_05_5002/man_05_5002.pdf

CPR 400.3.12. "Management of Classified Matter"

http://www-irn.sandia.gov/iss/CPR400.3.12/cpr400.3.12.htm

CPR 400.3.12.1 "Management of Classified Documents at SNL/NM and Remote Sites"

http://www-irn.sandia.gov/iss/CPR400.3.12.1/cpr400.3.12.1.htm

CPR 400.3.12.3 "Management of Accountable Classified Documents at SNL/NM and Remote Sites"

http://www-irn.sandia.gov/iss/CPR400.3.12.3/cpr400.3.12.3.htm

Management of classified matter at SNL/CA:

http://www.ran.sandia.gov/security/CPM/CPM2002/Index.htm

Classification Handbook for Derivative Classifiers at SNL

http://www-irn.sandia.gov/security/dept/classification/dchandbook/adcontent.htm

DOE Marking Handbook (March 2003):
http://www.wackenhut-oakridge.com/kw/CMPCDocuments/DOECmpcMarking/DOE%20CMPC%20Marking%20Handbook.pdf

The classified checklist in Appendix B includes some other requirements for classified reports.

Note: Refer to CPR 400.3.12.1, Preparing Secret or Confidential Drafts/Working Papers, for guidance on marking drafts.

QUICK REFERENCE GUIDE

A.4.1   Levels

All classified documents must be marked with the classification level. Refer to CPR 400.3.12.1 (section entitled Preparing Classified SAND Reports) for guidance:

http://www-irn.sandia.gov/iss/CPR400.3.12.1/cpr400.3.12.1.ac.htm#preparing_sand

Name of Notice	Instructions	Sample
Top Secret	Place at top and bottom on front of each page in bold letters at least one eighth-inch high. RD and FRD internal pages (including title page) are marked with **level and category**. The back of the last page is marked top and bottom with level and category. Cover and title page: If you have a Top Secret document, contact Classification Dept. (NM) or Communication Arts Dept. 8528 (CA) for marking requirements.	**TOP SECRET** For Distributions to locations outside the DOE Complex, the following is included on the title page and cover (CPR 400.12.3.1, Appendix C): **Dissemination and Reproduction Limitation – No further reproduction and/or dissemination of this document is permitted without the consent of the originator.** * *Exact wording may vary depending on classification circumstances. Contact Classification Department.

QUICK REFERENCE GUIDE

Name of Notice	Instructions	Sample
Secret	Place at the top and bottom on front of each page in bold letters at least one-eighth-inch high. RD and FRD internal pages (including title page) are marked with **level and category**. The back of the last page is marked top and bottom with level and category. Cover and title page:	**SECRET** For Distributions to locations outside the DOE Complex, the following is included on the title page and cover (CPR 400.12.3.1, Appendix C): **Dissemination and Reproduction Limitation – No further reproduction and/or dissemination of this document is permitted without the consent of the originator.*** *Exact wording may vary depending on classification circumstances. Contact Classification Department.

Confidential	Place at the top and bottom on front of each page in bold letters at least one-eighth-inch high. RD and FRD internal pages (including title page) are marked with **level and category**. The back of the last page is marked top and bottom with level and category.	**CONFIDENTIAL**
	Cover and title page:	For distributions to locations outside the DOE Complex: **Dissemination and Reproduction Limitation – No further reproduction and/or dissemination of this document is permitted without the consent of the originator.** * *Exact wording may vary depending on classification circumstances. Contact Classification Department.

A.4.2   *Categories*

In addition, various categories are required.

Name of Notice	Instructions	Sample
Restricted Data (RD) http://www-irn.sandia.gov/iss/CPR400.3.12.1/cpr400.3.12.1.htm	Cover/title page – top block under SAND No. Place on title page and cover, lower right corner. Level only on cover. Internal pages (including title page) carry **level and category**. Back of last page marked top and bottom with level and category.	**Restricted Data** **RESTRICTED DATA** **This document contains Restricted Data as defined in the Atomic Energy Act of 1954. Unauthorized disclosure subject to Administrative and Criminal Sanctions.** Note: Be sure to add the dissemination and reproduction statement associated with the level (Section A.4.1). **Classified by:** <Derivative classifier's name> **Title/Org.:** <Derivative classifier's title and organization> **Date: (optional)**<Date classification was determined> **Derived from:** <Designation of guide or source document and its date>

QUICK REFERENCE GUIDE

Name of Notice	Instructions	Sample
Formerly Restricted Data (FRD) http://www-irm.sandia.gov/iss/CPR400.3.12.1/cpr400.3.12.1.htm	Cover/title page – top block under SAND No. Place on title page and cover, lower left corner. Level only on cover. Internal pages (including title page) carry **level and category**. Back of last page marked top and bottom with level and category.	Formerly Restricted Data **FORMERLY RESTRICTED DATA** **Unauthorized disclosure subject to Administrative and Criminal Sanctions. Handle as Restricted Data in Foreign Dissemination, Section 144.b, Atomic Energy Act of 1954.** Note: Be sure to add the dissemination and reproduction statement associated with the level. **Classified by:** <Derivative classifier's name> **Title/Org.:** <Derivative classifier's title and organization> **Date:** (optional)<Date classification was determined> **Derived from:** <Designation of guide or source document and its date>

Name of Notice	Instructions	Sample
National Security Information (NSI) http://www-irn.sandia.gov/iss/CPR400.3.12.1/cpr400.3.12.1.htm	Cover/title page top block under SAND no. Cover page NSI documents dated after 4/1/97 are portion marked.	**National Security Information** Place on lower left side: **Classified by:** <Derivative classifier's name> **Title/Org:** <Derivative classifier's title/organization> **Date: (optional)** <Date classification was determined> **Derived from:** <Designation of guide or source document and its date> **Declassify on:** <Date, Event, or Exemption Category> **Note:** Be sure to add the dissemination and reproduction statement associated with the level (see A.3.1). **Ref.:** http://www-irn.sandia.gov/security/dept/classification/dchandbook/cover009.htm#markingsNSI

National Security Information (NSI) documents:

- Do not contain full category legend on the first page.
- Do not contain category marking on interior pages.
- Do include category marking on back of last page and back sheet.
- Do require caveat markings as applicable.

Name of Notice	Instructions	Sample
Naval Nuclear Propulsion Information http://www-irn.sandia.gov/security/cl assification/ucinnpi.htm	Refer to section A.2. May be classified or unclassified.	

A.5. Special Categories of Access

Refer to *Classification Handbook for Derivative Classifiers at SNL:* http://www-irn.sandia.gov/security/dept/classification/dchandbook/cover008.htm

C/UC*	Marking	Instruction	
C	Critical Nuclear Weapon Design Information (CNWDI)	Cover and title page, top block under SAND no. and NWD line When the document is to be (or may be) distributed to DoD, the above limitation should be placed under the NWD line. Cover and title page	**Critical Nuclear Weapon Design Information** **CRITICAL NUCLEAR WEAPON DESIGN INFORMATION** **DoD DIRECTIVE 5210.2 APPLIES**
C	FURTHER DISSEMINATION ONLY AS AUTHORIZED BY (GOVERNMENT AGENCY)	This notation applies to documents whose further dissemination within the receiving contractor facility is restricted to persons authorized by the addressee. Dissemination outside the facility is prohibited without the approval of the contracting activity.	

C/UC*	Marking	Instruction	
C	**Nuclear Weapon Data (NWD)** http://www-irn.sandia.gov/iss/CPR400.3.12.3/cpr400.3.12.3ac1.htm#preparing_reports See section on Preparing Sigma 14 SAND Reports.	Cover/title page – top block under SAND No. Note: A Sigma number must appear on all NWD reports. Cover and title page of Sigma 14 or 15: Cover and title page of Sigma 14 (not on Sigma 15); place above other legends: Check that all recipients are on the Sigma 14/15 access list maintained by Use Control Systems & Stockpile Support Dept. (NM) or International Security Dept. (CA). Applies to a minimum classification of Secret.	• **Nuclear Weapon Data** **Sigma ____** **SIGMA 14 (or 15)** **Sensitive Use Control** **Information** **Access Restricted** **This document may not be reproduced or disseminated beyond original distribution without approval of the originator, originating agency's Use Control Site Coordinator, or the NNSA HQ Use Control Site Coordinator.**
C	**REPRODUCTION REQUIRES APPROVAL OF ORIGINATOR**	This notation applies to documents that may not be reproduced without the specific, written approval of the originator.	

QUICK REFERENCE GUIDE

C/UC*	Marking	Instruction	
C/UCI	**PROGRAM DESIGNATED SPECIAL HANDLING** The Sandia Program manager must receive permission to use PDSH from the DOE Program Manager prior to publishing the document (DC Handbook). Requester must provide a letter of justification addressed to DOE OSTI from the agency or the Sandia/DOE Project Office that requires the distribution to be limited. The Technical Library will hold a copy of the letter for future inquiries or requests.	Cover/title page – top block under SAND No. (for UCI) Cover and title page Distribution list begins:	**Official Use Only / Program Designated Special Handling** **OFFICIAL USE ONLY / (Exemption Name) / PROGRAM DESIGNATED SPECIAL HANDLING** **In accordance with Program requirements, originating facility or DOE approval is required prior to further distribution.** **PROGRAM DESIGNATED SPECIAL HANDLING:**
C	**Foreign Government Information (FGI)** http://www-irn.sandia.gov/security/dept/classification/ucicfigi.htm	This notice is used on documents to ensure that information of foreign origin is not declassified prematurely or made accessible to nationals of a third country without the consent of the originator. This information is handled as if classified and Confidential, but with modified handling and protection requirements. For Confidential, use C/FGI-Mod cover sheet.	
U	**Internal Distribution Only (IDO)**	IDO is used for drafts and materials not going outside Sandia, or for corporate records. Insert Internal Distribution Only before the distribution list and on the top line of the cover and title page. IDO documents do not receive a SAND number.	
	NOFORN	Authorized only for intelligence information by authority of Director of Central Intelligence Information. May be used for NNPI (see DC Handbook).	

C/UC*	Marking	Instruction
SGI	**SAFEGUARDS INFORMATION** http://www-irn.sandia.gov/security/dept/classification/ucirsi.htm	This sensitive information pertains to the Nuclear Regulatory Commission. Mark pages containing SGI or all pages at the top and bottom with SAFEGUARDS INFORMATION If the document has multiple pages, the page marking must be placed at the top and bottom of: ■ the outside of the front and back covers, if any ■ the title page, if any ■ the first page of the text, if there is no front cover or title page ■ the outside of the back page, if there is no back cover Add **"Violation of protective requirements for Safeguards Information subject to civil and criminal penalties."** to the cover in the lower left corner. Mark front and back covers with the name, signature, title, and organization of the individual authorized to make an SGI determination and who has determined that the document contains SGI, and the date of the SGI determination in the lower right corner.

* Classified/Unclassified.

Distribution:

10   MS0619   Creative Arts, 3654
2   MS9018   Central Technical Files, 8944
2   MS0899   Technical Library, 4536